Big Cats in the Wild

A Visual Essay of Lions, Jaguars, Leopards, Pumas, and More

JOE MCDONALD

Amherst Media, Inc. ■ Buffalo, NY

Published by:
Amherst Media, Inc.
PO BOX 538
Buffalo, NY 14213
www.AmherstMedia.com

Publisher: Craig Alesse
Senior Editor/Production Manager: Michelle Perkins
Editors: Barbara A. Lynch-Johnt, Beth Alesse
Acquisitions Editor: Harvey Goldstein
Associate Publisher: Katie Kiss
Editorial Assistance from: Carey A. Miller, Roy Bakos, Jen Sexton-Riley, Rebecca Rudell
Business Manager: Sarah Loder
Marketing Associate: Tonya Flickinger

ISBN-13: 978-1-68203-324-1
Library of Congress Control Number: 2017949327
Printed in the United States of America
10 9 8 7 6 5 4 3 2 1

Notice of Disclaimer: The information contained in this book is based on the author's experience and opinions. The author and publisher will not be held liable for the use or misuse of the information in this book.

CONTENTS

ABOUT JOE AND MARY ANN MCDONALD

Joe and Mary Ann McDonald are professional wildlife photographers with a special interest in the Big Cats. They have photographed all seven of the Big Cats in a single year three different times—an adventure which took them to Africa, Asia, and South America. Joe is the author of eleven books on wildlife or wildlife photography, and Mary Ann is the author of twenty-nine children's books.

Joe and Mary spend over half the year in the field photographing, leading photo safaris, and conducting tours to worldwide destinations that typically include five continents. Learn more at www.hoothollow.com.

Photo by Joe McDonald.

Joe McDonald with a wild Cheetah that frequently scouted for game from the rooftops of safari vehicles in Kenya's Masai Mara Game Reserve. Photo by Mary Ann McDonald.

CONTRIBUTORS

Additional images were provided for this book by the following photographers:

Angus Fraser (pages 118–119, 121, and 125)
Cindy Marple (page 57)
Donna Salett (page 44)

INTRODUCTION

Mysterious, charismatic, majestic, powerful, frightening—these are just a few of the words that are often used to describe the world's Big Cats. Some, namely the Lion, Tiger, and Leopard, are virtually household names, known by nearly everyone. Others, like the Jaguar and Snow Leopard, are less familiar to some, but in their native lands they play important roles, functioning as keystone species that are critical to the health and stability of their environment. Nearly all, in today's world, face threats, with some of the most iconic, particularly the Tiger and Lion, facing extinction in the wild during the lifetime of many of you reading this sentence.

So, what defines a Big Cat? Technically, if we only consider a strict taxonomic classification, the Big Cat group might only include species in the *Panthera* genus: the Lion, Tiger, Jaguar, Leopard, and Snow Leopard. A related feline, the Clouded Leopard, is often included in this general classification, belonging to the same

The largest of the Big Cats is the Tiger. The Siberian or Amur Tiger can weigh over 700 pounds. Male Bengal Tigers, found in India, tip the scales at over 500 pounds. Male Siberian Tigers may be nearly 11 feet long, from nose to tail tip.

previous page, top—African Lions are the second-largest Big Cat, with males weighing 400 to 450 pounds. Females are much smaller and lighter, rarely weighing more than 300 pounds.

previous page, bottom—The size and weight of Jaguars varies considerably, depending upon the region. Jaguars in Central and northern South America are much smaller than those found in Brazil's Pantanal, where males may weigh over 250 pounds, making the Jaguar the third-largest Big Cat.

above—Pumas are known by many names and are commonly called Mountain Lions or Cougars in the United States and Canada. Pumas are the fourth-largest Big Cat. Males can reach 200 pounds or more.

taxonomic family *(Pantherinae)*, but Clouded Leopards are small, weighing no more than the largest of the Small Cats, the European Lynx. For that reason, we did not consider the Clouded Leopard as a Big Cat. Pumas and Cheetahs are not in the *Pantherinae* family, but both are big cats and equal or exceed the weight of the Snow Leopard. In fact, the Puma is the world's fourth-largest cat.

As a photographer, I hoped to one day photograph every one of the world's Big Cats, although I believed that doing so would be impossible. Fortunately, access and opportunities developed over the past several years to make that dream a reality, and my wife and I have succeeded in photographing all seven species in a single year, three different times. Our quest involved

travel to Africa, Asia, and South America, the three continents hosting all seven species. Ironically, one Big Cat, the Puma, is widespread in North, Central, and South America, but to have a realistic chance of even seeing this Big Cat, we had to travel to the southernmost regions of South America. The photographs in this book are a compilation from those quests and celebrate these beautiful and important predators.

. . . to have a realistic chance of even seeing this Big Cat, we had to travel to the southernmost regions of South America.

above—The fifth-largest Big Cat is the wide-ranging Leopard, found throughout much of Africa, parts of the Middle East, and central and southern Asia. Large males may top 200 pounds, although most Leopards weigh 150 pounds or less.

following page, top—In weight, the Cheetah nearly matches the Leopard, but its slender build creates the impression of a much lighter animal. A Cheetah's lithe build is designed for speed, not power and strength like the Leopard's.

following page, bottom—The smallest of the Big Cats is the elusive Snow Leopard, stalking the rugged mountain landscapes of central Asia. This incredibly camouflaged cat rarely weighs more than 120 pounds.

SECTION ONE

LIONS

Rocky Outcrop, Kenya (previous page, top)

From a rocky outcrop in Kenya's Masai Mara, a male African Lion surveys his territory, perhaps searching for his pride or on the lookout for trespassing males who might challenge him for possession of his kingdom.

Sound Asleep (previous page, bottom)

Like all cats, Lions spend a lot of their time sleeping, perhaps as much as twenty hours a day. Patiently waiting on a somnolent Lion will eventually be rewarded, but sadly, many tourists are too impatient and drive off to find more interesting sights.

Nocturnal Creatures (below)

Like all cats, Lions are nocturnal, and under the cover of darkness the cats hunt, play, and patrol their territory. In the course of a single night, Lions may travel a dozen miles or more, seeking game, marking territory, or driving off intruders.

Lions are nocturnal, and under the cover of darkness the cats hunt, play, and patrol their territory.

Mane Event (above)

Only male Lions have manes, and studies have shown that Lionesses prefer males with the darkest, thickest manes, as these are an indication of fitness and strength. Males reach their prime by age five, although a big black-maned lion like this one may be seven or eight years old.

Adaptation (following page)

In semi-desert habitats, the mane of a male Lion is sparse. In grasslands, a conspicuous, thick mane serves as a visual signal, alerting potential rivals of their presence, health, and vigor. In brush country, a conspicuous mane would have little value.

In the Dawn's Early Light (below)

At dawn, a male African Lion patrols his territory, roaring periodically to announce his presence. Lions also use scent to mark their territory, rubbing their chins on brush and spraying urine to warn off rivals.

Sounding Off (following page)

A lion's roar can be heard for miles. Safari guides say this roar appears to say, "Whose land? Whose land? Whose land? Mine! Mine! Mine!," and indeed the cadence of this roar mimics this phrase nicely.

Safari guides say this roar appears to say, "Whose land? Whose land? Whose land? Mine! Mine! Mine!"

A Mighty Roar (top)

Even a full-bellied Lion that is too lazy to stand can roar and will do so to answer that of another, sometimes prompting the entire pride to roar in a resounding chorus. Should you be fortunate enough to be in front of, and close to, a roaring Lion, you can actually feel the vibrations in the air.

Short-Lived Glory (bottom)

While one male Lion may possess a pride, his tenure may last for less than one year. Two Lions hold on to their pride on average for about two years, and three Lions an average of around three years, but coalitions of four or more male Lions can control a pride for several years.

I Stand Alone (top)

Lionesses often hunt alone, especially when the large herds of Gnus and Zebras have moved on. At various times, but especially at night, Lions may keep in contact with other pride members by roaring, although days may go by before members of a pride reunite.

Camouflage (bottom)

In tall grass, a Lioness hunkered down low can be nearly invisible, especially when the grasses dry and glaze a golden brown. Even in fresh green grass, a Lion is well camouflaged. Prey animals, like Gnus and Zebras, are color-blind, making a tawny cat in green grass virtually invisible as the two-color tonalities are similar.

A Warthog Escapes (left)

Although this Lioness waited until the Warthog passed only a few yards from where she lay in wait, the hunt was unsuccessful, as the Warthog sprinted away to safety.

Remains (top right)

Two Lionesses scramble to take possession of the remains of a Warthog they captured only a few minutes earlier, and in that time the carcass was torn in half. The violence and strength of several Lions fighting over their prey was shocking to see, giving us insight into the cats' incredible power.

Teamwork (bottom right)

When working together, several Lions can overcome and take down animals as large as a Hippopotamus or a sub-adult Elephant. A single Lion will prey upon baby Hippos and Elephants, although the mothers of either may deter any attempt at predation.

Zebra Fights Back (top)

Hunts can be dangerous, and a Lioness may be kicked by a Zebra, resulting in broken bones or teeth. Once, we found a starving Lioness who had her entire lower jaw removed by the well-placed kick of a fleeing Zebra.

In Pursuit (bottom)

A Lioness watches as her half-grown cub pursues an orphaned Gnu calf that trotted right into the pride. Lionesses train their cubs to hunt by capturing small prey that they'll release for their cubs to chase and tackle.

Gnu Story (top and center)

One of the funniest hunts I've ever photographed occurred when a Gnu, chased by a Lioness, suddenly spun and faced his attacker. The Lioness, programmed to chase prey, stopped and seemed confused. Apparently, so was the Gnu, as it soon turned and fled. The Lioness resumed the chase, but the Gnu did another face-off. This happened several times until the Lioness grew too tired to continue chasing the very brave Gnu.

A Distinct Disadvantage

(bottom)

A single Lioness is no match for an African Buffalo, nor is an entire pride when a Buffalo herd is present. These two Lionesses made no move toward the Buffalo, although their potential prey had other ideas.

The Lioness, programmed to chase prey, stopped and seemed confused.

Seeking Safety

(top and center)

A few minutes later, the Buffalo herd turned and approached the Lionesses, who retreated with their cubs. Several of the aggressive bulls charged the pride, scattering the Lions as they ran to the safety of a distant rock pile. The Buffalo would have killed any Lion cubs they caught.

A Group Effort (bottom)

Seven male Lions, all members of a large pride, piled onto a lone bull Buffalo that had wandered too close to the sleeping cats. Sometimes these struggles can last for hours, but the combined weight and strength of the seven male Lions ended this battle in only fifteen minutes.

Vulnerable (previous page, top)

At one point, a male Lion grabbed the snout of the bull, attempting to twist its neck and wrestle it to the ground. Head-to-head, a Lion is very vulnerable should the bull shake itself free, as the sharp horns of the Buffalo would skewer its attacker.

A Worthy Competitor (previous page, bottom left)

Although known as scavengers, Spotted Hyenas make many of their own kills, which are later stolen by Lions. A single Lion can drive off a pair of Hyenas, but often not without a fight.

Retribution (previous page, bottom-right images)

The tables turned after three Lionesses stole a Hyena's kill. The Hyenas whooped in reinforcements. In minutes, two dozen Hyenas appeared, chasing the Lionesses and biting any cat they could isolate. The battle ended when a male Lion arrived and drove off the Hyenas, who show far more fear or respect for the larger males.

All in the Family (above)

Lions are the only cat who live in large family groups, known as prides. Pride territories are inherited through the females, with male Lions keeping possession of the pride only until they are replaced by stronger males, who drive off or kill the previous "kings."

Not Welcome (bottom)

Lionesses, should they leave their own pride's territory, are not accepted by the females of another pride and will be killed. These females hope to bite the spine and paralyze or cripple the intruder.

Out of Harm's Way

(following page)

Young cubs will be carried by their mothers, sometimes for long distances. This Lioness was taking her cubs to a new den, fearful that the new males in the area would find and kill her cubs. Chances are, the males eventually did just that.

Newcomers (top)

New males are not immediately accepted by the Lionesses, as the newcomers will kill all of the young cubs, bringing females back into heat. To save their young, Lionesses will fight and, if the males are not tenacious, will drive the new males away.

Paternity (previous page, top left)

While male Lions will kill any young cubs they find when they take over a pride, they are tolerant and protective of their own cubs. Nonetheless, the first minutes after a Lioness introduces her cubs to the pride are tense, as cubs could be killed by mistake by their own father.

Orphan (previous page, top right)

We think this tiny cub was the orphan of a Lioness poisoned by herdsmen. Although carried to the pride by another Lioness, none of the females would nurse the young cub, despite the face that sisters and aunts normally nurse cubs that are not their own.

A Milestone (previous page, bottom)

After about ten days, a cub's eyes open, although the cubs are still nearly blind and very awkward. They'll remain at a den site for another six weeks before the Lioness takes them to join the pride.

Watchful (below)

A Lioness waits for one of her cubs as she begins the dangerous journey cross-country to join up with her pride. Cubs that can't keep up, perhaps by being weak or injured, may be abandoned during the trek.

Self-Protection (top and center)

A Lioness will be fiercely protective of her cubs, but in lean times a Lioness may eat everything she catches and let her cubs starve. While this seems cruel, a Lioness will have many opportunities during her lifetime to successfully raise cubs.

Sharing the Kill (bottom left)

Because a male's ownership of a pride is limited, a Lion may have only one chance to pass on his genes and have his cubs survive to maturity. Consequently, a male Lion will often share a kill with his cubs, but repel any Lionesses attempting to snatch a meal.

Caption This! (bottom right)

The camera captures a moment that begs for a caption. Is the cub complaining? Was there a bad joke? Bad breath? Or just the end of a pair of yawns?

Climbing (above, top and bottom)
Lions can climb trees and, if there are sturdy branches, even male Lions may climb high into a tree to sleep or to escape pesky biting flies. Like most cats, Lions cannot rotate their ankles as a squirrel (or a Clouded Leopard) can, and to descend, Lions either must jump or slide down a tree trunk backward.

Playtime (above, top, center, and bottom)
Cubs love to play, and often practice their later hunting skills, by biting the neck of their opponent. Play activity, regardless of the Lions' ages, is greatest a day or two after they have eaten, when hungers are sated and the cats have plenty of energy.

Adults Play, Too (top)

Even adult Lions will play. At first glance, the cats might appear to be fighting, but note that their claws are sheathed and their muzzles mask their canine fangs. In a real fight, both fangs and claws would be exposed—and used!

Charge! (bottom)

This charging Lioness's aggression was directed at her sister, whom she mauled right in front of me. Although I thought the cats were involved in a serious fight, when I reviewed my images later, I found that neither Lioness had her claws exposed.

Danger (following page, top)

All Big Cats are potentially dangerous, but in nearly thirty years, I've felt really threatened only once. A Lioness that completely ignored three other vehicles was fixated on mine, and every time we approached, the Lioness threatened to charge. Wisely, after recognizing this, we drove off.

On the Rise (following page, bottom)

At one time, Lions roamed across Africa, the Middle East, and western Asia into northern India. The Asian Lion is all but gone and is now restricted to the Gir Forest sanctuary and its surroundings in western India. Fortunately, this population, unlike those in most of Africa, is thriving and increasing in numbers.

SECTION TWO

TIGERS

Tigress (above)

The largest and arguably the most beautiful of all the Big Cats, a Tigress pauses on a dirt track in Bandhavgarh National Park in India. Poaching and habitat loss has severely impacted the Tiger, and today fewer than 3,000 survive in the wild.

Hidden (following page, top)

Tigers are forest, jungle, and high grassland animals, and seeing one in the wild is far more difficult than spotting a Lion in the African savannahs. Their vivid coloration and pattern actually forms an extremely effective camouflage, and when a Tiger steps into a forest or high grasses, it virtually disappears.

Obscured (following page, bottom)

A big male Tiger lies in the tall, dry grasses in Kazaranga National Park in northeastern India. I couldn't see it and had to have my guide aim my telephoto lens so that I could finally spot the cat. Can you see this Tiger?

Poaching and habitat loss has severely impacted the Tiger, and today fewer than 3,000 survive in the wild.

Claw Marks (top)

Deep claw marks high on a tree trunk help define a Tiger's territory. Male Tigers reach as high as they can to provide evidence of their size, warning rivals to beware.

I Was Here (bottom)

A young male Tiger pauses along the forest edge to scratch a tree trunk, creating a visual record of his passing. Territorial males will also scent-mark bushes and roar to broadcast their presence.

Tigers reach as high as they can to provide evidence of their size, warning rivals to beware.

A Fleeting Moment (above)

A large male Tiger pauses at the edge of a clearing. Because of the thick vegetation in their habitat, many Tiger sightings last only seconds. Surprisingly, although Langur Monkeys were in nearby trees, not a single monkey gave an alarm bark when this Tiger passed by.

Treasured Moments (top)

Sometimes you just get lucky and a Tiger appears, walking down a game trail or tourist track, oblivious of the vehicles or the tourists nearby. Not every game drive produces a Tiger, and every sighting should be treasured.

Tourists (bottom left)

Until recently, there were two ways to search for Tigers: driving the roads in open jeeps or bush-whacking on Elephant back. Today, nearly all Tiger tourism is done via jeeps on established tracks. Off-road driving is prohibited.

A Pleasant Surprise (bottom right)

A Tiger steps out of the forest and onto a game track, thrilling the photographers nearby.

The Approach (top left)

We had crested a small rise when this Tigress suddenly appeared, walking down the track toward us. To give the cat space, my driver put his jeep into reverse while I frantically shot some images.

B-2 (top right)

One of Bandhavgarh's most famous Tigers, B-2 was the son of a notorious Tiger named Charger, who got his name for his habit of false-charging jeeps. A year after this image was taken, old B-2 died.

B-2: A Second Sighting (bottom right)

On my first game drive in Tiger country, we spotted B-2 as he finished his meal, a Sambar Deer he had killed on the edge of a clearing. A few minutes later, B-2 walked by our jeep as he made his way to water.

Herd in Sight (top)

A Tigress stalks a distant herd of Spotted Deer. Like the Lion, a Tiger must be quite close before he charges if he is to be successful. Some Tigers, however, have learned to ambush deer at the edge of large lakes, driving them into the water where they are more easily caught.

Tiger vs. Porcupine

(center and bottom)

Tigers will eat a variety of prey, including the Indian Porcupine. When threatened, this large rodent will erect its quills, turn away from the threat, and charge backward in hopes of impaling its enemy. This Tiger seemed unaffected by the porcupine's quills.

Tigers will eat a variety of prey, including the Indian Porcupine.

The Spoils of Victory (top left)

A Tigress feeds upon another Tigress that she had killed after an earlier fight (she lost an eye during that conflict). Soon after taking this image, we were forced to leave so that a park supervisor could see this rare event for himself. We were furious!

Searching for Prey (bottom left)

At the edge of a forest pond a Tigress pauses, attempting to locate a Wild Boar rooting in the undergrowth. Minutes later, she crashed into the brush, and the loud squeals of the captured Boar announced the Tigress's success.

A Favorite Target (right)

Wild Boars are one of the Tiger's main prey. Because of the energy required in hunting, Tigers usually ignore small game, concentrating on Boars and various species of deer, all weighing over 50 pounds.

Wild Boars are one of the Tiger's main prey.

Tiger and Monkey (top)

A half-grown Tiger cub sprints across a road, carrying a Langur Monkey baby his mother captured only moments earlier. Because of their size, monkeys play a small role in a Tiger's diet.

Tiger and Fawn (center)

Photo credit: Donna Salett

After Donna Salett spent most of a morning waiting for a Tiger to appear, she heard brush cracking and a panicky squeal. A few minutes later, a Tiger carried a Spotted Deer fawn out of the forest.

Scent Marking (bottom)

Like all Big Cats, Tigers scent-mark their territory. This Tigress first rubbed her chin and face against the tree trunk, then turned and squirted urine before moving on.

An Accommodating Cat (following page, top)

While many Tiger encounters take place fleetingly and in deep forests, sometimes one gets lucky. This young male rested in the middle of a game track, tying up all vehicle traffic as everyone snapped photos of the accommodating cat.

At Risk (following page, bottom)

Young Tigers are at risk, as nomadic males will kill any cub they find. Although it was once thought that males played no role in Tiger family life, it is now known that resident males, owners of the territory, will share kills with the females and cubs and even play with their offspring.

Hydration (following pgae)

On hot days, Tigers frequently visit a waterhole in the late afternoon. Like all cats, Tigers scoop up water by curving their tongues, spatula-like, as they drink.

A Long Soak (top)

Unlike the Lion, which avoids getting its paws wet, Tigers love water and soak for hours on hot days when temperatures reach 110 degrees Fahrenheit. Many parks have created cement waterholes close to the tourist tracks to allow for easy viewing.

Backing Up (bottom)

A Tigress and her cubs soak in a pond in Tadoba National Park. Before immersing themselves, Tigers reverse direction and back into the water, settling their hindquarters first.

Tigers will roll over and immerse themselves completely, appearing to love their time in the water.

Snarl (top)
A cub snarls while soaking up to his neck. Tigers will roll over and immerse themselves completely, appearing to love their time in the water. This cub stayed in the water for several minutes after his two siblings and mother returned to the forest.

No Fear (center)
As the top predator in India's forests, Tigers have nothing to fear and will stretch out or sleep on exposed sand banks in the late afternoon or shaded glens at any time of day.

Fair Warning (bottom)
A film crew riding Elephants was following this Tiger, who was annoyed by the disturbance. Leaving the sanctuary of the forest, he snarled at my camera as he crossed a game track.

A Case of the Nerves (above)
In all my encounters with Tigers, I've only been nervous once, when a large male resting close to a game track grew annoyed as several jeeps stacked up nearby. I was watching from a distance but had to pass by to leave the park, so Mary Ann and I paused for a brief moment for a quick snap before we drove away.

In all my encounters with Tigers, I've only been nervous once, when a large male resting close to a game track grew annoyed as several jeeps stacked up nearby.

Conflict Avoidance (above)
Foresters travel through all the parks, either on foot or on bicycle. Although the rare Tiger attacks man, most glide off the road when they see anyone approaching.

Poaching (previous page)

Poaching, which is done to obtain the Tiger's beautiful skin and for body parts used in traditional medicines, threatens the existence of wild Tigers. Habitat loss, chiefly through deforestation, is an equally serious threat.

Dhole (top)

Perhaps the most efficient predator in the Indian forest is the Dhole, or Asian Wild Dog. Once quite common, sometimes roaming in packs that numbered in the dozens, large packs of Dholes occasionally cornered and killed Tigers. Today, it is rare to see packs larger than a family group of ten.

Sloth Bear (bottom)

Truly the most dangerous animal in the forest is the Sloth Bear, not the Tiger. When surprised, a Tiger usually runs off, but a Sloth Bear is far more likely to charge. Several people are seriously injured or killed each year by these animals.

McJaguar (top)

The Jaguar is the largest Big Cat in the Western Hemisphere and the most aquatic of all the Big Cats. This one, unmistakable with his distinctive right eye, was named "McJaguar" by the local river guides.

Tracks (bottom)

Tracks on a sandy beach mark the passage of a Jaguar. In contrast to many Big Cats, Jaguars in Brazil's Pantanal hunt at any time of day or night.

Lookout (top)

A Jaguar watches the river from a well-concealed lookout. In the Pantanal, Jaguars are now accustomed to seeing people. The cats are not shy, but spotting one can still be a challenge.

Jaguar at Rest (bottom)

Photo credit: Cindy Marple

My friend Cindy Marple took this photograph, the classic pose of a Jaguar overlooking a river while resting upon a log. I was at a different section of the river and missed seeing this cat and making my "dream shot."

Pantanal Cowboys (top)

The Pantanal is the world's-largest wetland, but the waters are seasonal, and during the dry months, much of the area is used for raising cattle. Before tourism demonstrated the economic importance of Jaguars, these cats were merely considered cattle killers, and often illegally shot or poisoned.

Tourism (bottom)

Jaguar tourism has increased steadily since its beginnings in 2007. Today, thousands of people visit the Pantanal in hopes of seeing Jaguars, and nearly everyone is successful. Most viewing is done from boats, and sometimes the rivers can get very crowded.

Jaguars patrol the river banks as they hunt for Caimans . . .

Hunting for Caimans (top)
Jaguars patrol the river banks as they hunt for Caimans, a type of crocodilian, and Capybaras, the world's largest rodent. Riverside viewing can be very rewarding as the cats travel along the shorelines, hunting for their favorite prey.

What's in a Name? (bottom)
The name "Jaguar" comes from an Indian word meaning "kills with a single leap." Although Jaguars love water, this one decided to jump across a small channel rather than wade across.

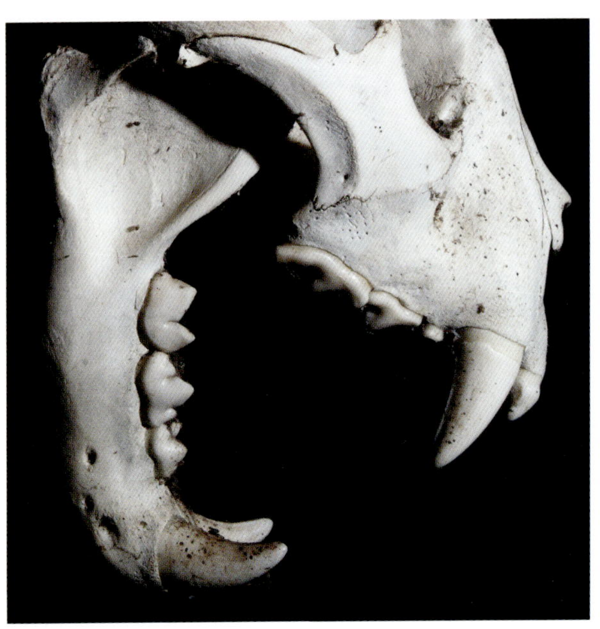

Jaguar Skull (top)

The Jaguar has the most powerful jaws of any of the Big Cats. When hunting Caimans, Jaguars bite through the thick, armor-like scales protecting the reptile's neck, instantly paralyzing the Caiman by doing so.

Yawn (bottom)

A yawning Jaguar reveals its powerful canines, or fangs. In some parts of the Amazon, Jaguars eat large river turtles, and older cat's fangs are worn down to thick stumps by their efforts.

Stealthy or Noisy (top)

A Jaguar slips through a clearing in the riverine forest. When hunting, a Jaguar moves silently, but when it is not, I've been amazed at how loud and noisy a Jaguar can be. This lack of stealth is testament to its position at the top of the food chain.

First Subject (center and bottom)

The first Jaguar I ever photographed required a careful approach. Cutting the motor, my guide slid out of the boat to wade, waist deep, as he pushed our craft up a shallow lagoon. Jaguars were shy and often retreated into the forest when seen. Today, a dozen or more boats with motors chugging loudly may keep pace with a Jaguar as it hunts the river's edge.

Eyes to the Camera (top)

Pausing during a long hunt, a male Jaguar stares fearlessly at the camera. A short time later, the cat stepped into the river, swimming directly to our boat as he headed toward the opposite shore. We quickly paddled out of his way.

Charge! (bottom)

Charging a group of Capybaras from forty yards, this Jaguar had little chance of catching one of these wary rodents. Successful hunts involve stalks in which the distance is only a few short yards.

Danger Detected (top left)

Although the Jaguar was completely invisible and still thirty yards away, this wary Capybara detected the danger and rocketed into the river for safety.

Alarm Call (bottom left)

When danger is spotted, Capybaras emit an alarm call, an explosive whistle-like snort that alerts others that a Jaguar is nearby. Despite this, other Capybaras often show no sign that they've received that message.

A Fortuitous Capture (top right)

Although the river offers the best escape route for a fleeing Capybara, it may not be enough. Seconds after a Capybara crashed into the river, it was followed by a Jaguar, who somehow captured the rodent underwater, dragging it to the shore and disappearing into the forest.

Night Vision (bottom right)

Like all cats, a Jaguar's nocturnal vision is improved by the tapetum, a light-reflecting layer of cells at the back of the eye that enhances light, much like a military night scope. A flash or spotlight aimed at the eyes reflects back, creating the eerie eye shine.

Hunter (top left)
A Jaguar intent on a Caiman stalks toward the camera, giving the impression that I'm the one being hunted. Like most hunts, this one ended unsuccessfully when the Caiman submerged.

Cooling Off (bottom left)
Jaguars readily take to the water. During a day's hunt, a Jaguar may cross a river several times or swim upstream or down for hundreds of yards. This is a behavior one doesn't expect to see in any cat.

Yacaré Caiman (top right)
The Yacaré Caiman is one of the Jaguar's main staples in the Pantanal. Growing to nearly eight feet long and equipped with powerful jaws, a Caiman is a potentially dangerous meal.

The Caiman Escapes (top)

A Jaguar charges across the shallows at a Caiman, who escaped. The success rate for all of the Big Cats is no greater than one in ten, or even worse. Many hunts, however, do not end with an actual chase, so the energy expenditure is not severe.

Carry Out (bottom)

Jaguars usually carry their kills into the forest, averaging nearly 70 yards from the point of capture. Large prey, like this Caiman stuck within a tangle of branches, may be consumed on the spot.

Mating Game (top and center)
A male Jaguar approaches a female on a sandy embankment. Only minutes earlier and in the shelter of the forest, the pair had mated, but now the female had a different mind-set. As the male approached, she charged.

Lover's Quarrel (bottom)
The female is to the left, and she means business. Her claws are out and she's ready to do damage. The male, in contrast, has his claws sheathed, and his snarl is one of displeasure and annoyance. After this spat, the two once again mated.

Courting (top)

A courting pair of Jaguars display an affectionate moment as the female butts heads with the larger male. Like most of the Big Cats, there is a significant difference in size between members of the two sexes.

Mom and Cub (bottom left)

A mother Jaguar scans the river while her half-grown cub looks on. Jaguar cubs will remain with their mother for nearly two years.

Flehmen (bottom right)

This odd-looking grimace identifies a behavior called flehmen, where the cat draws in particular scents. It is commonly seen when a Big Cat investigates a scent mark, or after a mother picks up her cub. Flehmen is not unique to the cat family; it is observed in Antelope and many hoofed animals as well.

Misconceptions (below)

The Jaguar's life history and behavior were poorly known before the Pantanal's population was observed, and many misconceptions occurred. Males are not as solitary and anti-social as once thought, and they sometimes share kills or spend time with other Jaguars of both sexes.

On the Beach (following page, top)

Two male Jaguars lounge together on a beach. Quite likely, these two adult Jaguars are brothers and know each other. They were seen together for several days before disappearing.

Quick Study (following page, bottom)

The two males study the opposite riverbank for potential prey before beginning their evening hunt. No one knows whether they will share a kill or if one will wait while the dominant cat feeds.

SECTION FOUR

LEOPARDS

Spot the Difference (top and center)

One of these cats is a Jaguar, the other a Leopard. Can you tell the difference? It is difficult in this image, but the Jaguar is shown in the second (center) image. Its rosette pattern of spots is larger, and you can see a few spots inside a few of the rosettes on its flank, the identifying mark of a Jaguar.

A Wide Range (bottom)

Leopards are the most wide-ranging of the Big Cats, found from southern Africa to southeastern Asia. They are most active at night, and in some cities in India they hunt the streets for dogs and pigs long after dark.

Camera Shy (top left)
The spotted coat of a Leopard provides extremely effective camouflage. Even in many national parks, Leopards can be shy and difficult to see.

Under Cover (top right)
A Leopard uses the cover of a fig tree to scan for prey or danger before crossing the Serengeti grasslands. Lions will kill Leopards if they have the chance, and Leopards will kill and eat Lion cubs if they have the opportunity.

In the Trees (bottom)
Leopards often climb trees where they rest, scan for prey, and keep out of the reach of Lions. Oddly enough, Leopards are on the lookout for Baboons, who will chase and kill any Leopard they see. If a Leopard spies a Baboon troop, the cat will leave the tree and seek better cover.

A Telltale Sign (top)

Sometimes the only evidence of a Leopard in the area is an old kill hanging in a tree. Leopards frequently carry their kills into trees, where they and their meal will be safe and out of reach of Lions and Spotted Hyenas.

A Favorite Haunt (bottom)

In Kenya's Samburu National Park, a female Leopard rests in a favorite tree. Some Leopards frequent particular trees or rocks, visiting them often, and spending the day there.

Vantage Point (following page)

Leopards are opportunistic hunters and will begin a hunt whenever game is spotted. While seeking game, Leopards may climb onto boulders or termite mounds, like this one, to achieve a better vantage point.

A Distinctive Roar (top)

A male Leopard gives its distinctive roar, a sound often described as a rough cough or a saw cutting through wood, a bit like this: "huh, huh, huh, huh-huh-huh."

Sanctuary (bottom)

Leopards are most frequently seen in or near thick cover, like brushy thickets or forests. In the grasslands of the Serengeti, trees are in short supply. Here, Leopards rely on the scattered yellow-barked acacia trees as sanctuaries, although cats occasionally are found miles from the safety of a tree.

A Mother's Motivation

(following page, top and bottom)

This female Leopard had a young cub hidden in a forest nearby. We were surprised to see her enter a Warthog burrow. Was she hunting for Warthogs, or was she looking for a new den to place her cub?

Paparazzi (top left)

With their beautiful spotted coats, combined with their often shy and elusive nature, Leopards are among the more sought-after animals by tourists. A posing Leopard always draws a crowd.

Too Close (top right)

This male Leopard snarled when a tourist vehicle approached too closely. Any sign of displeasure or annoyance should be heeded, as it is a sign the animal is stressed. Leopards, Lions, and Tigers have charged and entered vehicles when they've been pushed beyond their tolerance.

Slapstick (bottom)

This Leopard chose to flop onto its side just as my camera's shutter opened, capturing a funny, unlikely pose as the cat dropped to the ground.

In Miniature (top left)

Leopard cubs are miniatures of the adult, although their fur is a bit darker. Cubs are vulnerable to many predators, including Rock Pythons, Spotted Hyenas, Lions, and even male Leopards.

Affection (bottom left)

A Leopard cub and his mother rub their heads affectionately. Unlike other cats, Leopard cubs are quite independent and often wander far from their mother, sometimes for days at a time.

Chuffing (top right)

While resting on a termite mound, this mother Leopard chuffed a series of beckoning calls. A few minutes later, her large cub joined her, and they rubbed together for a short time before disappearing into the tall grasses.

While resting on a termite mound, this mother Leopard chuffed a series of beckoning calls.

Lunchtime (left)

I felt certain that this Vervet Monkey was safe when it took shelter in a twelve-foot high acacia tree, thick with branches and sharp thorns. I was wrong, as the Leopard leaped into the tree and caught the monkey, then carried it off to a hidden shelter.

A Healthy Appetite (right)

I've rarely seen a thin, unhealthy-looking Leopard, and there's a reason why: Leopards will eat just about anything, from insects to large Antelopes. This cub caught a rodent.

Large Prey (following page)

Leopards will take down surprisingly large prey, including big Antelope like Topi, Zebras, or, in this case, a Donkey that wandered away from its village. Preying on livestock, however, often leads to a Leopard's death.

Leopards will eat just about anything, from insects to large Antelopes.

Cat and Mouse (top)

After catching a baby Thompson's Gazelle, this Leopard began a game of cat-and-mouse, catching and releasing the baby several times.

About Face (bottom)

At one point, the baby Gazelle bravely turned and faced the Leopard, stopping the cat in its tracks. After regaining her confidence, the Leopard resumed her game, which ended badly for the baby.

Leftovers (following page, top)

If a Leopard feels safe, it will begin consuming its kill on the ground. Later, this Leopard carried the half-eaten baby Defassa's Waterbuck into a tree, where she fed upon the carcass for several days.

An Iron Jaw (following page, bottom left)

A Leopard's jaw strength must be incredible, as they use their jaws alone to carry prey items as large as a colt Zebra or half-grown Gnu into a tree.

Patience Pays Off

(following page, bottom right)

After locating a Leopard at dawn, we waited for five hours at a Gnu river-crossing point, hoping that the cat remained and would hunt if a crossing occurred. We were rewarded for our patience.

Sheer Will (top)

A half-grown Leopard cub tries unsuccessfully to drag the carcass of a ram Impala to a better location. This Impala was larger than the Leopard who killed it, but she was able to later drag that carcass into a nearby tree.

Impala (bottom)

A mother Leopard and her cub feed upon an Impala. A large kill will last a Leopard for several days, provided it is not discovered and stolen by Spotted Hyenas.

Formidable Prey (above)

Leopards often prey upon Warthogs, both adults and their young. Adult Warthogs are formidable, and we've seen Warthogs drive stalking Leopards into the safety of a tree. Amusingly, when the Warthogs turn and walk away, the Leopards climb down and resume the hunt, only to be chased back into the trees.

When Opportunity Knocks (top)

This Leopard holds a half-grown Warthog in her powerful jaws. Although a Leopard will feed upon a carcass until it is completely finished, should another animal wander by, she will hunt again.

Leopard vs. Dhole (bottom)

Minutes earlier, this Leopard won a dispute with an Indian Wild Dog, or Dhole, over possession of the kill. Earlier, the entire Dhole pack had stolen the kill, but when most of the pack had gone, the Leopard returned and carried his kill into the trees.

Dining on Deer (above)

In Pench National Park, India, a Leopard feeds upon a Spotted Deer in the safety of a rocky cliff where she can watch for Tigers. Tigers would steal the kill, but may kill the Leopard, too, thus eliminating potential competitors for prey.

SECTION FIVE

PUMAS

A Cat of Many Names (top)

Pumas go by many names, but throughout much of South America, they are known as Pumas. They are the fourth largest of all the cats; from head to tail, they can reach nearly eight feet. One, shot in Arizona, weighed over 275 pounds, the size of an African Lioness.

Torres del Paine National Park (bottom)

In the cliffs and caves of Torres del Paine National Park, Chile, Pumas seek shelter where they rest and spot for game. Although found in every country in both North and South America, excepting islands, this is the best location in the world to see and photograph Pumas.

Home Base (above)

A Puma pauses as her two young cubs play in the rocks below her. There are plenty of caves and crevices for shelter, and the area is a favorite denning site for mother Pumas.

In the Valley (top)

From a rocky outcropping, a Puma scans the sweeping valley below it, searching for its favorite prey, the Guanaco. Pumas have a varied diet that also includes birds, like the common Upland Goose, as well as Armadillos and Hognose Skunks.

Guanaco (bottom)

Framed by the backdrop of the rugged towers of Torres del Paine, a Guanaco stays on the alert for its only predator. If a Puma is spotted, the Guanaco gives a whinnying alarm whistle, alerting others of the danger.

Ambush (top)

Like almost all of the Big Cats, Pumas are ambush predators that wait until their prey is quite close before exploding into a charge. If cover is available, a Puma will wait for hours until a Guanaco walks near enough to make an attempt.

A Failed Attenpt (bottom)

A Guanaco walks by a mother Puma and her cub. Unfortunately the half-grown cub, still inexperienced in hunting, moved at the wrong time and the hunt was ruined. Pumas must hunt with surprise on their side.

Outrun (top and center)
This Puma had been feeding on a Guanaco killed days earlier when another Guanaco appeared on the horizon. The cat moved in and waited, and the Guanaco walked within a dozen yards of the crouching cat. Still, that distance was too far, and the Guanaco outran the Puma when it charged.

Protecting the Meal (bottom)
After feeding, Pumas will cover the carcass's remains with vegetation it scrapes free from the area nearby. Doing so shields the food from scavengers, especially Andean Condors that could strip the remaining meat free in several minutes.

A Rare Opportunity (top)

After finding this mother Puma and her cubs, we remained within sight of the cat the entire day, giving her time to become comfortable with our presence. Rarely does one have the chance to see, or photograph, a wild Puma with her cubs.

Kittens (bottom)

The Puma kittens spent much of their time on top of rocks where they played King of the Mountain or simply wrestled. As the top predator in southern South America, the cubs had little to fear.

Observer (top)

One of the Puma kittens studies us while keeping a rocky barrier between us for safety. When first born, Puma kittens have dark fur patterned with large spots that fade as the cub grows older.

The Chirp (bottom)

Pumas are distantly related to Cheetahs and like that spotted cat, Puma mothers emit a bird-like chirp to draw their cubs' attention. This high-pitched chirp carries surprising distances, even in the high winds common to this area.

A Playful Pair (top)

Two half-grown Puma cubs play on the edge of a small copse of trees. Play activity hones hunting skills, and Big Cats of all species practice neck holds and bites, moves they'll later employ to catch prey.

Premium Entertainment (bottom)

Playful cubs will entertain themselves with anything; their mother's tail, an old scrap of hide, a gnarly piece of driftwood, or a beckoning limb will do. This cub repeatedly leaped onto this branch, wrestling it to the ground each time.

Retractable Claws (top)

Pumas, like most cats, have retractable claws. When not in use, the claws are hidden in sheaths in their toe pads. By keeping the claws protected, they remain sharp and deadly, ready to be employed for gripping and holding on to prey.

In the Confines of a Cave (bottom left)

From the safety of a cave, a Puma watches the approach of another cat. Encounters with unrelated Pumas can be dangerous, and this Puma prudently chose to leave its shelter and seek safety elsewhere.

A Distinctive Appearance (bottom right)

Born without a tail, this young Puma was unique and instantly recognizable. When she matured, she disappeared, and it was feared she had been killed. Over a year later, she was seen again, in Argentina, in a protected area nearly fifty miles away.

A Popular Park (top)

Tourists from around the world visit Torres del Paine National Park to enjoy the scenery. Most are unaware that they share this landscape with the Puma, the animal with the greatest geographic range in the entire western hemisphere.

Persecuted Cats (bottom)

Outside the protection of the national park, Pumas are often persecuted as potential sheep killers. Fortunately, some landowners are now recognizing the tourist value Pumas have, and the future of the Puma here is brighter as tourism grows.

SECTION SIX

CHEETAHS

The Ultimate Sprinter

(previous page, top)

Built as the ultimate sprinter, the Cheetah is the fastest running land animal in the world. While their exact top speed is disputed, ranging from 55 to 75MPH, nothing is faster, and the acceleration is even more impressive—from 0 to 60MPH in 3 seconds.

Non-Retractable Claws

(previous page, bottom)

Unlike all other cats, Cheetahs have non-retractable claws, just like members of the dog family. The claws give better traction, just as track spikes would; consequently, they're not as sharp as other cats' and are not used for gripping prey.

The Spring in their Step

(top, center, and bottom)

Cheetahs have a flexible spine, providing for greater spring as their backbone flexes and extends, propelling the cat at tremendous speeds. The dew claws on their front legs function like hooks, snagging and tripping their prey.

The Tail (below)

The Cheetah's tail is used like a rudder, providing counter-balance as the cat tilts and swerves as it races after a fleeing gazelle, whose only hope of escape is by erratic moves left or right. In a straight-on race, the Gazelle has little chance.

In Close Proximity (following page, top)

Two five-month-old Cheetah cubs jog along, following their mother. Unlike Leopards, Cheetah moms keep their cubs nearby, only leaving them when she executes a hunt.

Wrestling (following page, bottom)

Wrestling plays an important role in their development, and cubs will practice tripping and take-downs soon after they begin to walk.

The Cheetah's tail is used like a rudder, providing counter-balance as the cat tilts and swerves . . .

An Opportunity for Fun (below)

Young cubs take every opportunity to play as they follow their mother. Cubs frequently jump at their mother's neck or tail, or practice neck holds on each other.

Dens (top)

Cheetah mothers give birth in rocky dens or isolated areas of tall grass. In the grasslands, cubs are killed when safari drivers go off-track, sometimes while they are looking for the cubs. Today, some parks close denning areas to prevent such deaths.

A Mother's Love (bottom)

Cheetah mothers will carry their very young cubs to new den sites. This cub is nearly too big to carry but had straggled along, requiring the attention of its impatient mother.

A Fairly Quiet Cat (top)

Like the Puma, a Cheetah's contact call, summoning a cub or another Cheetah, sounds like a chirping bird. Unlike other cats, Cheetahs are rarely vocal, and they, like Pumas, cannot roar.

Cub or Honey Badger? (center and bottom)

Cubs face many predators, especially in their first two months of life, when they are most vulnerable. Young cubs, however, resemble one of the most fearsome and aggressive African animals, the Honey Badger. From a distance, the two look quite similar, and this mimicry discourages any closer inspection.

Full of Affection (top)

Cheetahs are especially affectionate. This mother and her cub mutually groom, licking each other's face for several minutes. The Cheetahs' speed, beauty, and gentleness toward each other make them a favorite with many tourists on safari.

Guard Hairs (bottom)

This cub still sports the long guard hairs that give it a Honey Badger appearance. As the cub grows, only a scruff on the nape of the neck remains. It is often the only clue distinguishing a mother Cheetah from her nearly full-grown cubs.

A Tender Moment (top)
A cub pauses for a tender moment while her brother attempts to nurse. Cheetah mothers usually have three to five cubs, but their survival rate is low, and only 30 percent survive their first year. Only one in twenty cubs survive to adulthood.

Against the Odds (center)
This mother Cheetah in Kenya's Masai Mara raised five cubs to near adulthood. At least to this point, they avoided predation by Lions, Leopards, Spotted Hyenas, Pythons, and birds of prey. Still, their chance of survival was not guaranteed.

Together Forever (bottom)
Cheetah brothers remain together for life, while females are solitary unless they have cubs. Here, two adult male Cheetahs have cornered a mother Cheetah with her nearly full-grown cub. Although cubs can be killed in these encounters, this story ended happily.

An Elevated State (top)

Cheetahs often survey their hunting grounds from high vantage points, using rocks or even, in some parks, the roofs of safari vehicles. If prey is spotted, the cat will begin a careful stalk.

Striking Distance (bottom)

The Cheetah is a sprinter, not a long distance runner, and most chases do not begin until the Cheetah is within seventy yards or less from its potential victim. In this case, a Cheetah surprised a fawn that bolted in fear. Had the fawn lay still, as they often do, it would have been safe.

A Favorite Prey (top)
Fawn Thompson Gazelles are a favorite prey. When chased, their only hope of escape is to get lost in the fleeing herd, then drop down and freeze, motionless. Most don't do this, making them an easy catch for the Cheetah.

A Challenging Feat (center)
A Cheetah mother has a very difficult time hunting when she has cubs old enough to be mobile but still too young to understand the need to stay hidden from the eyes of their prey. Many hunts are ruined when curious cubs bound up to their stalking mother, alerting the prey.

After the Catch (bottom)
After catching this Steenbok, one of East Africa's smaller Antelopes, this Cheetah carried the kill until it reached a small thicket of brush. Cheetahs rarely feed in the open, as this may draw vultures who could in turn attract Hyenas and Lions that would steal the kill.

Hunting Practice (top)

Cheetah mothers will subdue but not kill small prey, like this Thompson Gazelle fawn, for their cubs to practice their hunting skills. The mother keeps a watchful eye, making sure that an inexperienced cub doesn't let a potential meal escape.

Rite of Passage (bottom)

When the cubs are old enough, the entire family may participate in a hunt. Here, three cubs joined their mother in a successful chase of a Reedbuck. At this stage, cubs are ready for independence.

The Cheetah and the Hare (top)

While stalking a distant herd of Gnus, this Cheetah surprised an African Hare. The Hare bounded away and the Cheetah, naturally, gave chase. The Hare got away.

A Surprising Turn of Events (bottom)

When a Cheetah charged into a line of Gnus, targeted on a calf, I was certain she would be successful. Surprisingly, the calf's mother did not join her herd in a bolting panic but turned and chased after her calf, driving the Cheetah off before she could make her kill.

Grooming (following page)

After a successful hunt, Cheetahs will fastidiously groom one another, licking blood clear of faces and necks. Unlike other cats, Cheetahs rarely, if ever, fight over a kill, and disputes over a meal involve little more than tugging matches.

Declining Numbers (top)

The five cubs shown earlier (“Against the Odds,” page 110) were reduced to three when we saw again almost a year later. Although the family successfully crossed the Talek River this time, a few weeks later, a Nile Crocodile snatched one of the cubs during a crossing.

Just Two Remain (bottom)

Still not independent, only two of the original five cubs were left. Considering one in twenty Cheetah cubs reach adulthood, we worried about the prospects of the two surviving cubs.

Strength in Numbers (above)

A lone male Cheetah has dim prospects for survival. If encountered by a pair or trio of males, he may be killed and even eaten. Male coalitions are comprised of brothers from the same litter, unlike male Lion coalitions, which can include unrelated but similar-aged cats.

A lone male Cheetah has dim prospects for survival. If encountered by a pair or trio of males, he may be killed and even eaten.

SECTION SEVEN
SNOW LEOPARDS

photo © Angus Fraser

An Elusive Cat (previous page, top)

Photo credit: Angus Fraser

The most elusive of the Big Cats, the Snow Leopard's exact numbers can only be estimated, ranging from 3,500 to 7,000 cats. Because of their remote and inhospitable range, it is even possible their numbers are much higher. No one knows.

In the High Country

(previous page, bottom)

Snow Leopards live in the high country, usually far above treeline, where, in winter, the land is enshrouded with snow. Cats follow ridgelines and valleys as they hunt for game and seek mates.

Cat Tracks (top)

The track of a Snow Leopard may be the only indication one is present, or had passed sometime during the night. In the mating season, their eerie yowls echo across the valleys, but for much of the year they are as silent as their mountain heights.

Limited Prey (bottom)

Prey diversity is limited in the high, dry elevations of the Himalayas. Marmots, Pika, Ibex, and several species of wild sheep comprise their diet. In winter, Snow Leopards follow the sheep herds, which seek open areas free of snow.

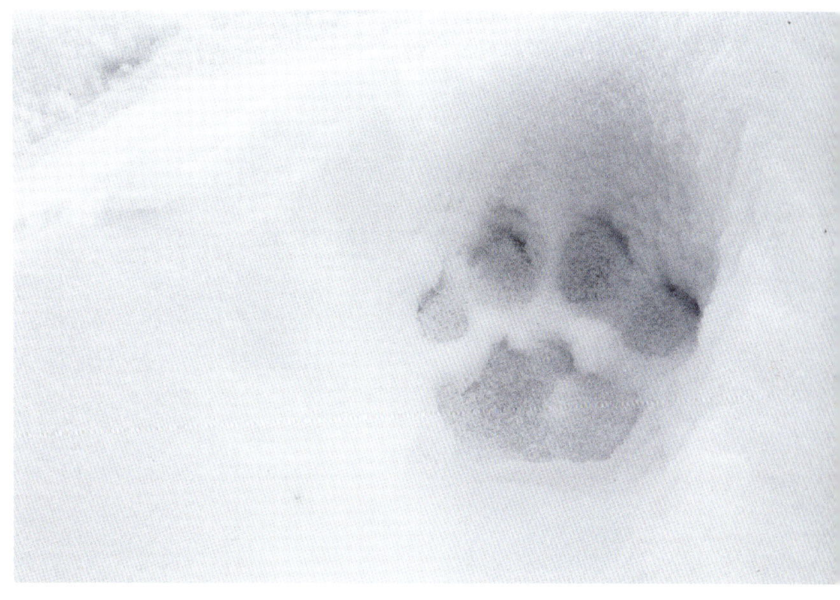

Blue Sheep (top)
Blue Sheep are a favorite prey, although their well-camouflaged coats make them difficult to spot from any distance. Snow Leopards must cover a lot of territory to encounter sheep and risk losing any kills they make to the Gray Wolves that also hunt this rugged landscape.

The Gray Ghost (center)
The Snow Leopard is known as "the Gray Ghost" for well-deserved reasons. Its camouflage is superb, perhaps the best of all the Big Cats, and one can disappear before your eyes when it moves into a rocky slope.

Patience is Key (bottom)
We waited for hours for this Snow Leopard to continue its stalk of an injured Blue Sheep as the sun disappeared behind the towering mountains. Cats have a greater chance of success after dark, but the outcome of this hunt remains unknown.

Small Herds (top)

Blue Sheep travel in small herds, providing protection to the group, with plenty of sharp eyes always on the alert. Herds move constantly as they seek the limited forage of this mountain desert.

Balance and Warmth (bottom)

The tail of a Snow Leopard is long and thickly furred and serves a vital role in maintaining balance when a cat bounds down or across a near vertical slope when in pursuit of sheep. On cold nights, the tail may also be used to cover the leopard's snout for warmth.

A Wide-Ranging Trek (above)

Photo credit: Angus Fraser

Snow Leopards range widely, and in summer may climb to 15,000 feet as they hunt Marmots and Sheep. In winter, marmots hibernate and sheep move down to lower elevations. The Leopards follow, although both remain high above treeline in most areas.

Snow Leopards range widely, and in summer may climb to 15,000 feet as they hunt Marmots and Sheep.

Shortly before dark, the Snow Leopard rose and walked toward us . . .

Mountain Gift (above)

We spent an entire day watching this Snow Leopard when it was only a shimmering lump of fur on a distant mountain slope. Shortly before dark, the Snow Leopard rose and walked toward us until only a steep gorge separated us, giving us a truly prized mountain gift.

INDEX